# ENERGY SECTOR STANDARD
# OF THE PEOPLE' S REPUBLIC OF CHINA
# 中华人民共和国能源行业标准

# Specification for Acceptance of Hydropower Projects

# 水电工程验收规程

**NB/T 35048-2015**

Replace DL/T 5123-2000

Chief Development Organization: China Renewable Energy Engineering Institute

Approval Department: National Energy Administration of the People's Republic of China

Implementation Date: September 1, 2015

China Water & Power Press

中国水利水电出版社

Beijing 2024

图书在版编目（CIP）数据
水电工程验收规程 : NB/T 35048-2015 = Specification for Acceptance of Hydropower Projects（NB/T 35048-2015） : 英文 / 国家能源局发布. -- 北京 : 中国水利水电出版社, 2024. 10.
ISBN 978-7-5226-2813-4
Ⅰ. TV512-65
中国国家版本馆CIP数据核字第2024GT7955号

ENERGY SECTOR STANDARD
OF THE PEOPLE'S REPUBLIC OF CHINA
中华人民共和国能源行业标准

**Specification for Acceptance of Hydropower Projects**

**水电工程验收规程**

**NB/T 35048-2015**

**Replace DL/T 5123-2000**

（英文版）

Issued by National Energy Administration of the People's Republic of China
国家能源局 发布
Translation organized by China Renewable Energy Engineering Institute
水电水利规划设计总院 组织翻译
Published by China Water & Power Press
中国水利水电出版社 出版发行
Tel: (+ 86 10) 68545888 68545874
sales@mwr.gov.cn
Account name: China Water & Power Press
Address: No.1, Yuyuantan Nanlu, Haidian District, Beijing 100038, China
http: //www.waterpub.com.cn
中国水利水电出版社微机排版中心 排版
北京中献拓方科技发展有限公司 印刷
184mm×260mm 16 开本 3 印张 95 千字
2024 年 10 月第 1 版 2024 年 10 月第 1 次印刷
**Price（定价）: ¥ 500.00**

# Introduction

This English version is one of China's energy sector standard series in English. Its translation was organized by China Renewable Energy Engineering Institute authorized by National Energy Administration of the People's Republic of China in compliance with relevant procedures and stipulations. This English version was issued by National Energy Administration of the People's Republic of China in Announcement [2023] No. 8 dated December 28, 2023.

This version was translated from the Chinese Standard NB/T 35048-2015, *Specification for Acceptance of Hydropower Projects*, published by China Electric Power Press. The copyright is reserved by National Energy Administration of the People's Republic of China. In the event of any discrepancy in the implementation, the Chinese version shall prevail.

Many thanks go to the staff from the relevant standard development organizations and those who have provided generous assistance in the translation and review process.

For further improvement of the English version, any comments and suggestions are welcome and should be addressed to:

China Renewable Energy Engineering Institute
No. 2 Beixiaojie, Liupukang, Xicheng District, Beijing 100120, China
Website: www.creei.cn

Translating organization:

China Renewable Energy Engineering Institute

Translating staff:

| | | | |
|---|---|---|---|
| HE Wei | CHENG Zhengfei | CHENG Li | LI Xiang |
| ZHUANG Hongzhi | LIU Biao | HE Jialong | HOU Shaokang |
| YUE Lei | | | |

Review panel members:

| | |
|---|---|
| GUO Jie | POWERCHINA Beijing Engineering Corporation Limited |
| QIE Chunsheng | Senior English Translator |
| YAN Wenjun | Army Academy of Armored Forces, PLA |
| LI Zhongjie | POWERCHINA Northwest Engineering Corporation Limited |

| | |
|---|---|
| CHEN Lei | POWERCHINA Zhongnan Engineering Corporation Limited |
| YE Bin | POWERCHINA Huadong Engineering Corporation Limited |
| JIANG Jinzhang | POWERCHINA Huadong Engineering Corporation Limited |
| QI Wen | POWERCHINA Beijing Engineering Corporation Limited |
| WEI Fang | China Renewable Energy Engineering Institute |
| CHE Zhenying | IBF Technologies Corporation Limited |


National Energy Administration of the People's Republic of China

# 翻译出版说明

本译本为国家能源局委托水电水利规划设计总院按照有关程序和规定，统一组织翻译的能源行业标准英文版系列译本之一。2023 年 12 月 28 日，国家能源局以 2023 年第 8 号公告予以公布。

本译本是根据中国电力出版社出版的《水电工程验收规程》NB/T 35048—2015 翻译的，著作权归国家能源局所有。在使用过程中，如出现异议，以中文版为准。

本译本在翻译和审核过程中，本标准编制单位及编制组有关成员给予了积极协助。

为不断提高本译本的质量，欢迎使用者提出意见和建议，并反馈给水电水利规划设计总院。

地址：北京市西城区六铺炕北小街 2 号
邮编：100120
网址：www.creei.cn

本译本翻译单位：水电水利规划设计总院

本译本翻译人员：何　伟　程正飞　程　立　李　祥　庄洪志　刘　彪　何佳龙　侯少康　岳　蕾

本译本审核人员：

郭　洁　中国电建集团北京勘测设计研究院有限公司

郄春生　英语高级翻译

闫文军　中国人民解放军陆军装甲兵学院

李仲杰　中国电建集团西北勘测设计研究院有限公司

陈　蕾　中国电建集团中南勘测设计研究院有限公司

叶　彬　中国电建集团华东勘测设计研究院有限公司

江金章　中国电建集团华东勘测设计研究院有限公司

齐　文　中国电建集团北京勘测设计研究院有限公司

魏　芳　水电水利规划设计总院

车振英　一百分信息技术有限公司

国家能源局

# Announcement of National Energy Administration of the People's Republic of China [2015] No. 3

According to the requirements of Document GNJKJ [2009] No. 52, "Notice on Releasing the Energy Sector Standardization Administration Regulations (*tentative*) and detailed implementation rules issued by National Energy Administration of the People's Republic of China", 203 sector standards such as *Carbon Steel and Low Alloy Steel for Pressurized Water Reactor Nuclear Power Plants—Part 31: 15Mn Forgings for Containment Vessel*, including 106 energy standards (NB) and 97 electric power standards (DL), are issued by National Energy Administration of the People's Republic of China after due review and approval.

Attachment: Directory of Sector Standards

National Energy Administration of the People's Republic of China

April 2, 2015

Attachment:

**Directory of Sector Standards**

| Serial number | Standard No. | Title | Replaced standard No. | Adopted international standard No. | Approval date | Implementation date |
|---|---|---|---|---|---|---|
| … | | | | | | |
| 69 | NB/T 35048-2015 | Specification for Acceptance of Hydropower Projects | DL/T 5123-2000 | | 2015-04-02 | 2015-09-01 |
| … | | | | | | |

# Foreword

According to the requirements of Document GNKJ [2010] No. 320 issued by National Energy Administration of the People's Republic of China, "Notice on Releasing the Development and Revision Plan of the First Batch of Energy Sector Standards in 2010", and after extensive investigation and research, summarization of practical experience, and wide solicitation of opinions, the drafting group has prepared this specification.

The main technical contents of this specification include: river closure acceptance, reservoir impoundment acceptance, turbine-generator unit start-up acceptance, special works acceptance, hydropower complex acceptance, and completion acceptance.

The main technical contents revised are as follows:

—Revising the application scope of this specification.

—Revising the content of stage acceptance and completion acceptance.

—Revising the requirements for acceptance basis.

—Revising the requirements for acceptance organization.

—Revising the requirements for acceptance application.

—Revising the conditions that shall be met for the river closure acceptance, impoundment acceptance, turbine-generator unit start-up acceptance, special works acceptance, hydropower complex acceptance, and completion acceptance.

—Adding the requirements for the procedures and methods for the river closure acceptance, impoundment acceptance, turbine-generator unit start-up acceptance, special works acceptance, hydropower complex acceptance, and completion acceptance.

—Revising the requirements for the acceptance of special works.

—Adding the preparation requirements for the acceptance program and acceptance certificate.

—Revising the list of relevant documents and information for acceptance.

National Energy Administration of the People's Republic of China is in charge of the administration of this specification. China Renewable Energy Engineering Institute has proposed this specification and is responsible for its routine management. Energy Sector Standardization Technical Committee on Hydropower Investigation and Design is responsible for the explanation of specific technical contents. Comments and suggestions in the implementation of this specification should be addressed to:

China Renewable Energy Engineering Institute
No. 2 Beixiaojie, Liupukang, Xicheng District, Beijing 100120, China

Chief development organization:

China Renewable Energy Engineering Institute

Participating development organization:

Power Construction Corporation of China

Chief drafting staff:

| | | | |
|---|---|---|---|
| WEI Zhiyuan | ZHENG Xingang | LI Sheng | CHEN Huiming |
| ZHAO Quansheng | WANG Runling | SUN Baoping | YU Qinggui |
| WANG Yuansheng | YANG Hua | LIN Zhaohui | LI Xiushu |
| FANG Hui | WEI Xiaowan | MENG Fanfan | FENG Haibo |
| WAN Wengong | | | |

Review panel members:

| | | | |
|---|---|---|---|
| WANG Minhao | ZHOU Jianping | QIN Wangyu | WANG Youcai |
| HONG Yunbo | PENG Tubiao | LI Shisheng | ZHAO Kun |
| YANG Zhigang | BIAN Wei | XIE Honglin | YAN Jun |
| LYU Mingzhi | WANG Renkun | ZHANG Zongliang | CHEN Yinqi |
| ZHOU Chuiyi | PAN Jiangyang | SHI Qingchun | YAN Lei |

# Contents

# 1 General Provisions

**1.0.1** This specification is formulated in accordance with the relevant state laws and regulations, with a view to standardizing and strengthening the acceptance management of hydropower projects to ensure the safety of the project and the life and property upstream and downstream, meet the requirements of project tasks, and achieve project functions.

**1.0.2** This specification is applicable to the acceptance of national key construction hydropower projects and hydropower projects approved by national authorities. The start-up acceptance of other large-sized hydropower plants (with a total installed capacity of 300 MW or above) shall also be carried out in accordance with this specification. This specification may be applicable to the acceptance of other hydropower projects not mentioned above.

**1.0.3** The acceptance of hydropower projects includes stage acceptance and completion acceptance:

**1** Stage acceptance, which includes the river closure acceptance, impoundment acceptance, and turbine-generator unit start-up acceptance. Resettlement acceptance shall be conducted before the river closure acceptance and the impoundment acceptance.

**2** Completion acceptance, which shall be conducted after passing the hydropower complex acceptance, resettlement acceptance, environmental protection acceptance, soil and water conservation acceptance, fire control acceptance, occupational health and safety acceptance, project final accounting and archives acceptance, as well as special works acceptance. The acceptance of hydropower complex, etc. is not restricted by the acceptance of special works.

**1.0.4** For hydropower projects, acceptance must be conducted in time before the river closure, impoundment, and turbine-generator unit start-up and after the project completion.

**1.0.5** The national energy authority is responsible for the oversight of the acceptance of hydropower projects.

The local energy authorities are responsible for and participate in the administration, guidance, coordination and oversight of the acceptance of hydropower projects in their respective administrative areas according to their terms of reference.

**1.0.6** An acceptance committee shall be established to organize the river closure acceptance, impoundment acceptance, turbine-generator unit start-up acceptance, hydropower complex acceptance, special works acceptance, and completion acceptance. The acceptance shall be conducted in a scientific, objective, fair and standardized manner.

**1.0.7** The acceptance of hydropower projects shall be carried out on the basis of the following documents:

**1** Relevant laws, regulations and sector regulations of China.

**2** Relevant national and sector specifications, codes and technical standards.

**3** Project approval documents.

**4** The approved project feasibility study report (or original preliminary design); special reports, major design change report, budgetary estimate adjustment, corresponding review and approval comments, and relevant design documents and drawings approved by the competent department of the project through technical review.

**5** Contract documents related to project construction and applicable quality standards and technical documents specified in the contract.

**1.0.8** The project owner is responsible for the resettlement acceptance, environmental protection acceptance, soil and water conservation acceptance, fire control acceptance, occupational health and safety acceptance, project final accounting and archive acceptance in accordance with relevant laws and regulations.

**1.0.9** The acceptance costs for hydropower projects are included in the project cost estimates.

**1.0.10** In addition to this specification, the acceptance of hydropower projects shall comply with other current relevant standards of China.

# 2 Basic Requirement

## 2.1 Acceptance Organization

**2.1.1** The provincial energy authority is responsible for the impoundment acceptance, hydropower complex acceptance and completion acceptance, entrusting a qualified organization as the technical presiding organization to establish an acceptance committee and invite relevant departments, parent firm of the project owner and relevant organizations and experts to participate in the above acceptance. The provincial energy authority serves as chairman of the acceptance committee, or the technical presiding organization may be entrusted to serve as chairman of the acceptance committee. The provincial development and reform commission, the technical presiding organization and the parent firm of the project owner serve as deputy chairmen of the acceptance committee.

For a project constructed on an inter-province (autonomous region or municipality directly under the central government) river, the provincial energy authorities concerned are jointly responsible for the impoundment acceptance, hydropower complex acceptance and completion acceptance.

**2.1.2** For the river closure acceptance, the project owner organizes the acceptance committee jointly with the provincial development and reform commission and the provincial energy authority, and invites relevant departments, parent firm of the project owner, and relevant organizations and experts to participate in the acceptance. The project owner serves as chairman of the acceptance committee, and the provincial development and reform commission, the provincial energy authority and parent firm of the project owner serve as deputy chairmen of the acceptance committee.

**2.1.3** For the turbine-generator unit start-up acceptance, the project owner and the power grid operator jointly organize the acceptance committee and invite the provincial development and reform commission, the provincial energy authority, relevant departments, the parent firm of the project owner, and relevant organizations and experts to participate in the acceptance. The project owner and the power grid operator determine the chairman and deputy chairmen of the acceptance committee through consultation, but the provincial development and reform commission, the provincial energy authority, and the delegated office of the national energy authority shall serve as deputy chairmen of the acceptance committee.

The turbine-generator unit start-up acceptance for large hydropower plants in other projects may be organized in accordance with relevant sector regulations, but the provincial development and reform commission, the provincial energy authority, the delegated office of national energy authority and the power grid operator shall serve as deputy chairmen of the acceptance committee.

**2.1.4** For the special works acceptance, the completion acceptance presiding organization establishes the acceptance committee, and when necessary, may collaborate with relevant departments or organizations to jointly establish such a committee.

**2.1.5** An expert team is set up under the acceptance committee, which is organized by the technical presiding organization to undertake the technical acceptance and the work authorized by the acceptance committee.

**2.1.6** The project owner is responsible for organizing relevant preparations before acceptance and assists the acceptance committee in the acceptance work.

## 2.2 Acceptance Application

**2.2.1** For the river closure acceptance, the project owner shall apply to the provincial energy authority six months before the planned closure.

**2.2.2** For the impoundment acceptance, the project owner shall apply to the provincial energy authority six months before the planned impoundment according to the project schedule, and copy the application to the technical presiding organization simultaneously.

**2.2.3** For the turbine-generator unit start-up acceptance, the project owner shall apply to the provincial energy authority three months before the first unit start-up acceptance, and copy the application to the power grid operator simultaneously.

**2.2.4** For the hydropower complex acceptance, the project owner shall apply to the provincial energy authority three months before the planned hydropower complex acceptance according to the project schedule, and copy the application to the technical presiding organization simultaneously.

**2.2.5** For the special works acceptance, the project owner shall apply to the provincial energy authority three months before the planned acceptance according to the project schedule, and copy the application to the technical presiding organization simultaneously.

**2.2.6** For the completion acceptance, the project owner shall, within 12 months after the project is substantially completed or all turbine-generator units are put into operation, carry out the work concerned and implement the resettlement acceptance, environmental protection acceptance, soil and water conservation acceptance, fire control acceptance, occupational health and safety acceptance, project final accounting and archive acceptance in accordance with relevant laws and regulations. The project owner may apply for the completion acceptance separately or together with the hydropower complex acceptance. If submitted separately, the completion acceptance application shall be submitted to the provincial energy authority, and copied to the technical presiding organization simultaneously.

## 2.3 Main Workflow of Acceptance Committee

**2.3.1** After receiving the acceptance application, the acceptance presiding organization shall establish an acceptance committee jointly with the departments and organizations concerned through consultation.

**2.3.2** The acceptance committee shall formulate and issue the acceptance program. See Appendix A of this specification for the acceptance program.

**2.3.3** The technical presiding organization shall organize an expert team to carry out on-site inspection and technical acceptance of the project, listen to the report of relevant organizations and review the documents and information for acceptance.

**2.3.4** The acceptance committee holds a project acceptance meeting to study and confirm the acceptance conditions.

**2.3.5** Acceptance results are proposed.

## 2.4 Acceptance Result

**2.4.1** The main project acceptance result refers to the acceptance certificate. The acceptance certificate shall be signed by the chairman and deputy chairmen of the acceptance committee, accompanied by the list of signatures of all members of the acceptance committee and the list of signatures of representatives of the organization applying for acceptance. The comments made by the expert team of the acceptance committee on technical acceptance shall be attached to the acceptance certificate. See Appendix B of this specification for the acceptance certificate.

**2.4.2** The acceptance conclusion must be agreed upon by more than two-thirds of the members of the acceptance committee. Any member of the acceptance committee who disagrees with the acceptance conclusion shall clearly document the comments and affix the signature on the acceptance certificate.

**2.4.3** In the case of any disputes during the acceptance, the chairman of the acceptance committee shall coordinate and resolve the disputes, and include the comments on major issues put forward by the committee members in the memorandum attached to the acceptance certificate. If the chairman's decision is opposed by more than half of the committee members or in the case of major issues difficult to decide, the acceptance committee shall report to the acceptance presiding organization for ruling, and the ruling document shall be attached to the acceptance certificate; major issues shall be reported to the provincial energy authority in time.

**2.4.4** In the turbine-generator unit start-up acceptance of other large hydropower plants (with a total installed capacity of 300 MW or above), if there is any discrepancy between the project owner and the power grid operator, the discrepancies shall be timely reported to the provincial energy authority for coordination.

**2.4.5** The acceptance certificate shall be made in octuplicate and sent to relevant organizations by the acceptance presiding organization. The xeroxed certificate shall not be used as the original.

## 2.5 Documents for Acceptance

**2.5.1** The project owner is responsible for organizing and coordinating all relevant organizations to prepare the documents required for acceptance within the given time limit. Each organization shall be responsible for the authenticity of the submitted documents. See Appendix C of this specification for the list of documents required for acceptance and the list of documents for future reference.

**2.5.2** The originals of the documents for acceptance shall be stamped with the official seals of document providers.

**2.5.3** The main documents for acceptance shall be archived as required after acceptance.

# 3 River Closure Acceptance

**3.0.1** The river closure of a hydropower project is characterized by cutting off the main river channel, retaining water by the cofferdam for main works, and diverting river water through diversion structures. The river closure acceptance shall perform the inspection and acceptance of diversion structures, closure preparation, cofferdam design and construction scheme according to the approved design scheme, to ensure the safety of project construction and the safety of life and property upstream and downstream.

## 3.1 Conditions for Acceptance

**3.1.1** The diversion release structures related to river closure have been basically completed according to the design requirements and are qualified to allow water passing, and the uncompleted works construction shall not be affected by diversion.

**3.1.2** The underwater concealed works related to closure in the main works have been completed, and the quality meets the criteria specified in the contract documents.

**3.1.3** The closure implementation scheme and cofferdam design, and the construction scheme have been approved by the review organized by the project owner. Various preparations have been made according to the approved closure implementation scheme, including the organization, manning, machinery, roads, material preparation, communications, and emergency response. The construction schedule of the project after closure has been implemented, and the physical progress before flood season meet the requirements of flood control.

**3.1.4** The flood control scheme after river closure has been approved, its measures have been basically implemented, and the upstream flood forecasting has been arranged, which have met the requirements of flood control.

**3.1.5** The resettlement works before river closure determined in the planning and design of resettlement has been completed, and the competent resettlement department of the provincial government in the area where the project is located has organized and completed the stage acceptance, made comments on acceptance, and reached a clear conclusion that the river closure will not be affected.

**3.1.6** The temporary navigation or transfer transportation issues of navigable rivers have been basically solved, or an agreement has been reached with relevant departments.

**3.1.7** The safety appraisal of special works has been completed for diversion structures before river closure; for the hydropower projects with staged diversion by riverbed, the safety appraisal of (closure) special works have been completed for some permanent structures (including concealed works) used for temporary water retaining before water flowing into the foundation pit, and the definite conclusion for operation has been reached.

**3.1.8** The report on quality supervision at the closure stage has been submitted, and a conclusion that the works quality meets the requirements of river closure has been reached.

## 3.2 Main Acceptance Activities

**3.2.1** Check and confirm on site that the diversion release structures related to the closure of the main riverbed have been completed; assess the water passing conditions of diversion structures according to the check results of structural safety and discharge capacity of diversion structures; check and assess the construction quality of diversion structures and defect treatment results; check and assess the feasibility of the scheme of removing the cofferdams or rock ridges at the entrance and exit of diversion structures.

**3.2.2** Check and assess on site the construction quality of the underwater part of main works related to river closure.

**3.2.3** Check and assess the feasibility of the cofferdam design and construction scheme.

**3.2.4** Check and assess the flood control measures during the construction period of the cofferdam and main works.

**3.2.5** Check and assess on site the feasibility of the closure scheme approved by the project owner.

**3.2.6** Check and assess on site the preparations for river closure and cofferdam construction.

**3.2.7** Check the implementation of reservoir special works and resettlement before river closure, the technical documents for acceptance of resettlement during river closure, and the implementation of the comments on pending issues in the stage acceptance of resettlement.

**3.2.8** Check the documents and information for acceptance submitted by the project participants.

## 3.3 Procedures, Methods, and Results for Acceptance

**3.3.1** The acceptance presiding organization shall organize the preparation and issue of the acceptance program, and put forward the acceptance plan and acceptance preparation requirements.

**3.3.2** The acceptance presiding organization shall organize an acceptance expert team to inspect the actual situation of the project on site, understand the river closure conditions and preparations, check and assess the cofferdam design scheme and flood control scheme after river closure, conduct technical acceptance of river closure, and make comments on the acceptance of river closure.

**3.3.3** The acceptance committee shall hold an acceptance meeting for river closure on the basis that the acceptance of the resettlement work at the river closure stage has been completed and the acceptance requirements of river closure have been satisfied.

**3.3.4** The river closure acceptance meeting include the following:

**1** Listen to the reports from the parties involving project development, design, supervision and construction, the comments on the quality supervision and conclusions of the safety appraisal of (closure) special works, the comments on resettlement acceptance at the closure stage, and the comments from the expert team.

**2** Check the river closure scheme and the preparations.

**3** Check the flood control standard and measures for the project safety after river closure.

**4** Review relevant documents and confirm all the conditions for the river closure acceptance.

**5** Give suggestions on dealing with the pending issues related to acceptance.

**6** Discuss and complete the river closure acceptance certificate.

**3.3.5** The acceptance presiding organization shall submit the acceptance certificate to the provincial energy authority, and copy to relevant organizations.

# 4 Reservoir Impoundment Acceptance

**4.0.1** The reservoir impoundment of hydropower projects is characterized by cutting off the water flow through the diversion structures, retaining water by dams and starting impoundment in the reservoir. The reservoir impoundment acceptance shall perform the acceptance of the water retaining and water discharge structures related to the reservoir impoundment and near-dam reservoir banks according to the approved project design scheme to ensure the safety of the hydropower projects and the safety of life and property upstream and downstream after the reservoir impoundment.

## 4.1 Conditions for Acceptance

**4.1.1** The physical progress of the dam foundation and its seepage control works, dam and other water retaining structures, joint grouting of dams and seepage control works of the reservoir basin has met the requirements of reservoir impoundment (to the target water level) of the project, the quality of works is qualified, and the project construction and flood control will not be affected by the impoundment.

**4.1.2** The inlet and outlet gates and trash racks of water conveyance structures related to impoundment have already been installed in place and ready for water retaining.

**4.1.3** The water release structures that need to be put into operation after the impoundment have been substantially completed, the gates and hoists required for impoundment and discharge have been installed, equipped with reliable power supply, and can operate normally.

**4.1.4** The internal and external monitoring instruments and equipment of each structure have been embedded and tested according to the design requirements, and their initial values have been obtained.

**4.1.5** The unstable reservoir bank and reservoir leakage affecting the safe operation of the project after the impoundment have been treated according to the design requirements, and the reservoir-induced earthquake monitoring facilities have been installed according to the design requirements, and their background values have been obtained.

**4.1.6** The plugging gates, gate slots, and hoisting equipment of diversion release structures are checked and in good condition, and can meet the requirements of impoundment plugging.

**4.1.7** The impoundment planning scheme and construction planning have been prepared and passed the review organized by the project owner; necessary preparations have been made for impoundment, including organization, staffing, roads, communications, leakage plugging and emergency response.

**4.1.8** The regulations on the reservoir operation and the power plant operation dispatching, and the flood control scheme for initial operation after impoundment have been formulated and reviewed or approved by the competent departments; navigation and downstream water supply during impoundment have been realized; the hydrologic telemetry and forecasting system can meet the impoundment requirements of the project.

**4.1.9** The environmental protection works and soil and water conservation works affected by the impoundment have been substantially completed, not affecting construction progress after the impoundment. The ecological flow discharge scheme has been finalized and measures have been put in place, and arrangements have been made for the various aspects downstream that might be affected by ecological discharge.

**4.1.10** The operator has got ready for impoundment, has been staffed with qualified personnel, and has developed operating procedures for all control equipment; various facilities have met the requirements of initial operation.

**4.1.11** The corresponding special works in the reservoir area affected by the impoundment have been substantially completed, and the relocation and reservoir area clearance has been completed. The provincial resettlement authority has organized and completed the stage acceptance of the resettlement, and issued acceptance opinions, which gives a clear conclusion that the resettlement will not affect the impoundment.

**4.1.12** The report on safety appraisal of the impoundment has been submitted, which gives a clear conclusion that it is feasible to implement the impoundment.

**4.1.13** The report on the quality supervision at the impoundment stage has been submitted, which gives a conclusion that the works quality meets the impoundment requirements.

**4.1.14** The emergency response plan for the impoundment period has been formulated and registered.

## 4.2 Main Acceptance Activities

**4.2.1** Check on site whether the physical progress meets the impoundment conditions.

**4.2.2** Based on monitoring data analysis results, assess the impoundment and dam safety according to the approved design scheme, the physical progress of the project and the changes of the implementation scheme, as well as the review results of the stability and structural safety of water retaining structures and water release structures.

**4.2.3** Check and assess the feasibility of the flood control standard and measures for the initial operation after the impoundment.

**4.2.4** Check and assess the feasibility of the impoundment planning scheme.

**4.2.5** Check and assess the navigation and downstream water supply (including ecological flow) schemes and preparations for the initial operation after the impoundment.

**4.2.6** Check and assess the plugging construction scheme and preparations of diversion structures. Check that the diversion release structures, plugging gates, plugging sections and water retaining tunnel sections are in the condition to bear the maximum water pressure during impoundment after the gate is closed.

**4.2.7** Check and assess the construction quality of the water retaining and release structures and defect remedy results according to the self-inspection results of the construction contractor, and on-site inspection results of the supervisor and the third party.

**4.2.8** Check and assess the review results of the stability of the dam and near-dam reservoir bank slope according to the geological review investigation results.

**4.2.9** Check the installation and testing of gates and hoists of water retaining and release structures and the preparations of plugging gates of diversion structures.

**4.2.10** Check the completion conditions of the dam safety monitoring works, check the monitoring of initial operation and the acquisition of initial values, and assess the monitoring data analysis results.

**4.2.11** Check the implementation of the comments on pending issues in the safety appraisal of the impoundment.

**4.2.12** Check the implementation of reservoir special works, resettlement and the reservoir basin clearance, the technical documents for resettlement acceptance during impoundment, and the implementation of the comments on pending issues in the stage acceptance of resettlement.

**4.2.13** Check the documents and information for acceptance submitted by the project participants.

## 4.3 Acceptance Procedures, Methods and Results

**4.3.1** The presiding organization organizes the preparation and issue of the acceptance program which proposes the acceptance plan and requirements for preparatory work.

**4.3.2** The acceptance presiding organization organizes an acceptance expert team to perform on-site inspection and technical acceptance. The expert team shall inspect the actual situation of the project on site, analyze and confirm the impoundment conditions that the project shall have, check the acceptance documents submitted by each project participant, and give suggestions on dealing with the problems identified in the impoundment acceptance.

**4.3.3** After the resettlement acceptance at impoundment stage has completed and a conclusion has been made that the requirements of impoundment have been satisfied, the acceptance presiding organization shall organize the expert team to check the rectification and implementation of the on-site inspection comments, check the technical documents related to acceptance, and present expert team opinion on impoundment acceptance.

**4.3.4** After the technical acceptance has completed by the acceptance expert team and a conclusion has been made that the impoundment acceptance requirements are satisfied, the acceptance committee holds a meeting for impoundment acceptance.

**4.3.5** The impoundment acceptance meeting includes the following:

**1** Listen to the reports from the parties involving project development, design, supervision, construction, equipment manufacturing and installation, operation, etc., the comments of quality supervision and the conclusions of safety appraisal of impoundment, and the comments on resettlement acceptance at the impoundment stage, and the comments from the acceptance expert team.

**2** Check the impoundment scheme and the preparations.

**3** Check the flood control standard and flood control measures after impoundment.

**4** Review relevant documents and confirm all the conditions for the impoundment acceptance.

**5** Give suggestions on dealing with pending issues.

**6** Discuss and complete the impoundment acceptance certificate.

**4.3.6** The acceptance presiding organization submits the acceptance certificate to the provincial energy authority and copies to the relevant organizations.

**4.3.7** The acceptance committee may authorize the acceptance expert team on its behalf to conduct the acceptance for the hydropower project whose staged impoundment does not involve corresponding (staged) resettlement acceptance, otherwise, the hydropower project must be subject to acceptance of the acceptance committee.

**4.3.8** After the completion of impoundment acceptance, the project owner shall report the relevant conditions during and after impoundment to the presiding organization within the specified time. The presiding organization shall organize the expert team to check the impoundment on site as appropriate.

**4.3.9** The presiding organization shall report the relevant conditions during and after impoundment to the provincial energy authority as per the regulations.

# 5 Turbine-Generator Unit Start-up Acceptance

**5.0.1** The turbine-generator unit start-up acceptance is the preliminary assessment and comprehensive quality evaluation on the manufacture and installation of the water conveyance system, and turbine-generator units and auxiliaries after the units and associated electromechanical equipment have been installed and test-qualified and before they are put into initial commercial operation. Each turbine-generator unit shall be subject to trial operation and start-up acceptance before the initial operation to ensure that the unit is commissioned in a safe and reliable way as a whole. The units can be grid-connected only after they are test-qualified and accepted for handover.

## 5.1 Conditions for Acceptance

**5.1.1** The hydropower complex has passed the impoundment acceptance, and the physical progress of the project meets the requirements of initial power generation; the water conveyance system has been completed in line with design documents and up to specific standards; the reservoir level has reached the minimum operating level for power generation after impoundment; the tailrace outlet has been cleared as required; the special safety appraisal report on the water conveyance system has been submitted, with a conclusion that the system meets the conditions of water filling for trial operation.

**5.1.2** The inlet and outlet gates of the water conveyance system and their hoists for the turbine-generator units for acceptance have been installed, tested and are ready for operation; the inlets and outlets of other water conveyance systems without turbine-generator units have been reliably sealed.

**5.1.3** The civil works in the powerhouse have been substantially completed according to the contract documents and design drawings, the section with the turbine-generator units for acceptance has been fenced and isolated, and the passages on each floor and the lamps in the plant have been put in place, which can ensure the safe construction of the project and the safe trial operation of the turbine-generator units for acceptance; the drainage system in the plant has been installed, tested and are ready for operation; the flood control and drainage facilities in the plant area have been put in place to ensure safe operation in flood season.

**5.1.4** The turbine-generator units for acceptance and their auxiliaries, including oil, compressed air and water systems, have been installed and have completed the testing and trial run, and their quality meets the requirements specified in the contract documents; the utility system and automation system of the whole plant have been put into operation, meeting the needs of the trial operation of the turbine-generator units for acceptance.

**5.1.5** The primary and secondary electrical equipment for the turbine-generator units is test-qualified and works accurately and reliably, meeting the requirements of step-up, power transformation, power transmission, measurement, control and protection; the earthing resistance of the earthing system of the whole plant meets the design requirements; the computer supervisory control system has been installed and test-qualified.

**5.1.6** The system communication, intraplant communication system and external communication system have been established as designed, and have been installed and test-qualified.

**5.1.7** The civil works of booster station, switchyard, outgoing line yard, and other parts have been completed according to the design requirements, and the direct lightning flash protection system has been established, meeting the safe transmission of high-voltage electrical equipment; the necessary power transmission lines have been built, and their relay protection equipment has been installed and test-qualified.

**5.1.8** Firefighting facilities meet the requirements.

**5.1.9** The instruments and equipment provided by the installation and testing unit meet the needs of the trial operation of turbine-generator units. The operator of the hydropower station has been mobilized, the production and operation staffing meet the needs of the initial commercial operation of the turbine-generator units, the operating procedures have been formulated, and the relevant instruments and equipment meet the needs of the initial commercial operation of the turbine-generator units.

**5.1.10** The quality supervision report for the turbine-generator unit start-up acceptance has been submitted, with a conclusion that the works quality meets the start-up acceptance requirements of turbine-generator units.

## 5.2 Main Acceptance Activities

**5.2.1** Review and approve the start-up trial operation program and trial operation plan of turbine-generator units.

**5.2.2** Check and assess whether the physical progress and quality of the turbine-generator units and their auxiliaries, common equipment for hydraulic machinery, electrical equipment, hydraulic steel structures and hoists and hydraulic structures of the water conveyance system meet the requirements specified in the contract documents and meet the start-up requirements of the turbine-generator units.

**5.2.3** Check, through on-site inspection and review of documents, whether all the conditions required in Section 5.1 and other conditions considered necessary by the committee are met, put forward the activities that must be completed and precautions before the unit start-up, conclude whether the unit is approved for start-up trial operation, and determine the time for the initial start-up.

**5.2.4** Assess the inspection results of commissioning test of the turbine-generator unit for acceptance, and confirm whether conventional turbine-generator units meet the conditions of 72 h on-load trial operation or whether reversible units meet the conditions of 15 d on-load trial operation.

**5.2.5** Confirm the list of project items for acceptance and handover.

## 5.3 Acceptance Procedures, Methods and Results

**5.3.1** The presiding organization organizes the preparation and issue of the acceptance program which proposes the acceptance plan and requirements for preparatory work.

**5.3.2** After the turbine-generator unit to be accepted has been installed and test-qualified, the acceptance presiding organization shall organize the acceptance expert team to inspect and assess the manufacture, construction, installation and testing of water conveyance systems of turbine-generator units, turbine-generator units and their auxiliaries, utility systems such as oil, air and water, and equipment and facilities relating to the turbine-generator unit for acceptance, such as electrical equipment, and control and protection equipment, put forward the problems and give suggestions on their solutions. According to the inspection results, the acceptance expert team shall draw a conclusion whether the turbine-generator unit is ready for trial operation, and determine the start-up time of the turbine-generator unit; put forward the preparations that must be completed and precautions before the turbine-generator unit is started.

**5.3.3** After all items specified in the trial operation of turbine-generator units are tested, the acceptance expert team shall confirm the list of project items for acceptance and handover jointly with the construction contractor, operator and erector.

**5.3.4** According to the acceptance plan and the site situation, after the start-up trial operation is completed for turbine-generator units and when the conditions of the turbine-generator unit start-up acceptance specified in Section 5.1 are met, the acceptance committee shall hold the meeting for the turbine-generator unit start-up acceptance and complete the following:

1 Listen to the reports on the quality inspection by supervisors, the

acceptance reports from the parties involving project development, design, construction, quality supervision, safety monitoring, operation, and fire control as well as preparations for trial operation and acceptance and handover.

**2** Assess the results of the commissioning test; review and confirm whether conventional units meet the conditions of 72 h on-load trial operation or the conditions of 15 d on-load trial operation for reversible units.

**3** Authorize the acceptance expert team to review the report on 72 h on-load trial operation of conventional units or 15 d on-load trial operation of reversible units.

**4** Draft the appraisal of the turbine-generator unit start-up acceptance, conclude whether the turbine-generator unit passes the trial operation examination and whether it is feasible to put the turbine-generator unit into initial commercial operation.

**5** Give suggestions on dealing with the non-compliant items and existent problems.

**5.3.5** The project owner shall make the report on 72 h on-load trail operation of conventional units or 15 d on-load trial operation of reversible units, and the acceptance expert team shall review the report and give comments.

**5.3.6** The chairman and deputy chairmen of the acceptance committee officially issue the turbine-generator unit start-up acceptance certificate with test records attached according to the review comments of the acceptance expert team.

# 6 Special Works Acceptance

**6.0.1** For special works such as intake and navigation structures that have independent functions and render unique benefits and roles, the completion acceptance presiding organization shall organize the special works acceptance accordingly if the special works need separate acceptance due to being put into operation earlier or later than the overall project plan.

**6.0.2** For a postponed or suspended special works that does not affect the main function of the hydropower complex, the completion acceptance presiding organization may organize the acceptance of the special works upon its completion after the hydropower complex acceptance.

## 6.1 Conditions for Acceptance

**6.1.1** The special works has been substantially completed according to the requirements of contract documents and design drawings, the project quality is up to the design standard, and the construction site has been cleared as required.

**6.1.2** The special works has been put into trial operation and has met the approved functional requirements.

**6.1.3** The equipment manufacture and installation for special works have been proved safe and reliable through testing and trial operation and have met the requirements of the contract documents and design.

**6.1.4** The observation instruments and equipment have been embedded according to the design requirements, and the initial values have been obtained; a perfect initial operation monitoring and data processing management system has been established, and complete initial operation monitoring data and analysis reports are available.

**6.1.5** The project quality events have been properly treated and the quality defects have been substantially remedied to ensure the normal operation of project; the winding-up works and the remedy of defects have been completed by the construction contractor within the defects notification period.

**6.1.6** The project operator has got ready for handover and operation.

**6.1.7** The safety appraisal report with an affirmative conclusion on safe operation for the special works has been presented.

**6.1.8** The quality supervision report for special works with an affirmative conclusion on the acceptance of the special works has been presented.

## 6.2 Main Acceptance Activities

**6.2.1** Check the construction status of the special works in accordance with the approved design documents, including the physical progress and functions, the implementation of major design changes, and the project costs.

**6.2.2** Check the scheme modifications and design review during implementation.

**6.2.3** Check and assess the functions and structural safety of structures of the special works according to their trial operation, the treatment of problems identified during the trial operation, the design review and the analysis of safety monitoring data.

**6.2.4** Check and assess the function and safety of hydraulic steel structures and electromechanical engineering in special works according to the trial operation of hydraulic steel structures and electromechanical equipment for the special works, and the treatment of problems identified during the trial operation.

**6.2.5** Check the quality of the special works in the construction and equipment installation, and remedy of defects identified during the trial operation, and assess the works safety and quality.

**6.2.6** Check the documents and information for the acceptance.

**6.2.7** Put forward comments and requirements to deal with the pending issues.

## 6.3 Acceptance Procedures, Methods and Results

**6.3.1** The acceptance presiding organization prepares and issues the acceptance program which proposes the acceptance work plan and preparation requirements.

**6.3.2** Upon completing the safety appraisal of special works, the acceptance presiding organization delegates the acceptance expert team to initiate on-site inspections. The acceptance expert team shall examine project construction and operation, investigate and analyze the acceptance conditions that shall be met, check the acceptance documents proposed by the project owner and operator, and propose rectification comments and improvement requirements for the non-compliant issues.

**6.3.3** Based on the on-site inspection results, the acceptance expert team may call for a third-party check, test, inspection and supplementary demonstration for the defects and pending issues.

**6.3.4** After rectifications are completed by the project owner and operator according to the on-site inspection comments, the acceptance presiding organization delegates the acceptance expert team to review the on-site rectification, check the technical documents for acceptance, and make comments on the special works acceptance.

**6.3.5** When the acceptance expert team completes the technical acceptance, and the acceptance requirements for the special works are met, the acceptance committee convenes the acceptance meeting for the special works.

**6.3.6** The special works acceptance meeting includes the following:

**1** Listen to the reports from the parties involving project development, design, safety monitoring, operation, quality supervision, etc., the comments on completion safety appraisal of special works and technical acceptance of expert team, review relevant acceptance documents, and discuss whether the conditions for the special works acceptance are met.

**2** Check the construction and operation of special works according to the approved project tasks, assess the quality, safety and initial operation, and give suggestions on dealing with pending issues.

**3** Discuss and complete the special works acceptance certificate.

**6.3.7** The acceptance presiding organization shall submit the acceptance certificate to the provincial energy authority, and copy to relevant organizations.

# 7 Hydropower Complex Acceptance

**7.0.1** The hydropower complex acceptance refers to the special acceptance of the project functions and structure safety at the completion stage according to approved project tasks after the project has been completed as per the approved design scale and criteria and after the verification by its initial operation within a specified period.

**7.0.2** The hydropower complex acceptance includes the acceptance of all project structures for water retaining, water release, water conveyance and power generation, navigation, fish pass, etc., and their hydraulic steel structure works, safety monitoring works, electromechanical works, as well as permanent slope works and near-dam reservoir bank slope treatment works. Except for the intake and navigation structures which can operate and produce benefits separately and may be subject to acceptance as special works, all the related project structures and electromechanical equipment installation works approved in the project feasibility study report and required to be completed in the same period shall be included in the hydropower complex acceptance.

## 7.1 Conditions for Acceptance

**7.1.1** The hydropower complex has been completed according to the approved design scale and criteria, and the project quality is up to standard.

**7.1.2** Major project design changes have gone through confirmation formalities.

**7.1.3** The winding-up works have been completed and the quality defects have been remedied.

**7.1.4** The initial operation of the project has experienced at least one flood period, the multiyear regulation reservoir needs to experience at least two flood periods, and the highest reservoir level has reached or basically reached the normal pool level.

**7.1.5** All turbine-generator units have been able to operate normally at the rated output for at least 2000 hours (including standby hours for a pumped storage power station).

**7.1.6** Except for special works, each works operates normally and meets the requirements of corresponding design functions.

**7.1.7** The safety appraisal agency has provided the completion safety appraisal report with the conclusion of safe operation.

**7.1.8** The quality supervision report for the completion stage has been submitted, which gives a conclusion that the project quality meets the completion acceptance requirements.

## 7.2 Main Acceptance Activities

**7.2.1** Check the basic information on the project complex construction according to the approved design documents, including the physical progress, the project tasks and the functions, the procedures for major design changes, the quantities and the project costs.

**7.2.2** Check and assess the safety of water retaining structures based on the analysis of safety monitoring data of their initial operation after the impoundment, the review of stability and structural safety, and the treatment of problems found.

**7.2.3** For large-sized projects, check the functions of flood discharge and energy dissipation structures, and assess the safety and reliability of flood control based on the comparative analysis of the hydraulic model test results according to their operation status and treatment of problems identified in the initial operation after impoundment.

**7.2.4** Check and assess the functions and safety of the structures of the water conveyance and power generation system according to their initial operation situation and the analysis of initial operation safety monitoring data after the impoundment, and the treatment of the problems identified in the inspection by water releasing.

**7.2.5** Check and assess the stability of the permanent slope works of the project structures and the near-dam reservoir bank slope treatment works according to the analysis of the initial operation safety monitoring data and the stability review after the impoundment, and the treatment of the problems identified.

**7.2.6** Check and assess the effective operation of the project safety monitoring system according to the initial operation and maintenance of the project safety monitoring system after impoundment, instruments and equipment conditions, and the processing and analysis of monitoring data.

**7.2.7** Check and assess the operational functionality and reliability of the turbine-generator units according to the results of trial operation, load tests and maintenance of the units.

**7.2.8** Check the construction quality of civil works and equipment installation, and remedy of defects identified in the initial operation of the project after impoundment to assess the project quality.

**7.2.9** Check the conclusion of the completion safety appraisal and the implementation of pending issues.

**7.2.10** Check the project acceptance documents.

**7.2.11** Put forward suggestions and requirements for the treatment of the pending issues.

## 7.3 Acceptance Procedures, Methods and Results

**7.3.1** The presiding organization organizes the preparation and issue of the acceptance program which proposes the acceptance plan and requirements for preparatory work.

**7.3.2** The presiding organization organizes the acceptance expert team to carry out on-site inspection and technical acceptance. The acceptance expert team shall check the actual situation of construction and initial operation of the project on the site, analyze the acceptance conditions required for the project, check the acceptance documents submitted by the project owner and conduct technical acceptance, and give suggestions on dealing with the non-compliant issues and propose the requirements for the work to be supplemented and improved.

**7.3.3** Based on the on-site inspection results, the acceptance expert team may call for a third-party check, test, inspection and supplementary demonstration for the defects and pending issues.

**7.3.4** After rectifications are completed by the project owner according to the comments of the on-site inspection, the presiding organization for acceptance organizes the acceptance expert team to review the on-site rectification, check the technical documents related to acceptance and make comments on the hydropower complex acceptance.

**7.3.5** When the technical acceptance has been completed by the acceptance expert team and the acceptance requirements are met, the acceptance committee holds the special acceptance meeting.

**7.3.6** The hydropower complex acceptance meeting includes the following:

1 Listen to the reports from the parties involving project development, design, safety monitoring, operation, etc., the comments of quality supervision and the conclusions of safety appraisal of completion, the technical acceptance comments from the acceptance expert team, and review relevant acceptance documents, and discuss whether the acceptance conditions for hydropower complex are met.

2 Check the construction and operation of the hydropower complex

according to the approved project tasks, assess the quality, safety and initial operation of the project, and give suggestions on dealing with the pending issues.

**3** Discuss and complete the hydropower complex acceptance certificate.

**7.3.7** The presiding organization shall submit the acceptance certificate to the provincial energy authority, and copy to relevant organizations.

# 8 Completion Acceptance

**8.0.1** The completion acceptance refers to the overall acceptance of a hydropower project after it has been built according to the approved design documents and all special acceptances at the completion stage have been completed.

## 8.1 Conditions for Acceptance

**8.1.1** The hydropower complex acceptance, resettlement acceptance, environmental protection acceptance, soil and water conservation acceptance, fire control acceptance, occupational health and safety acceptance, project final accounting and archive acceptance have been performed in accordance with relevant state laws and regulations, and there are definite written conclusions for passing acceptance.

**8.1.2** The special works which has not been subject to acceptance will not affect the safety of the hydropower project and the life and property upstream and downstream, and the construction and completion acceptance plan of this special works has been worked out.

**8.1.3** The pending issues in the completion acceptance have been properly solved, and the winding-up works have been completed.

**8.1.4** Other relevant specifications and regulations are met.

## 8.2 Summary Report of Completion Acceptance

**8.2.1** The project owner summarizes the hydropower complex acceptance, resettlement acceptance, environmental protection acceptance, soil and water conservation acceptance, fire control acceptance, occupational health and safety acceptance, project final accounting and archive acceptance, and provides a summary report on the completion acceptance.

**8.2.2** The summary report of the completion acceptance shall include the following:

**1** Project overview.

**2** Briefing of each acceptance before completion acceptance.

**3** Main conclusions of acceptance certificates (enclosing all completion acceptance certificates).

**4** Treatment of the problems and suggestions described in acceptance certificates.

**5** Acceptance situation of the special works completed after hydropower complex acceptance, or the work schedule for the special works not subject to acceptance at the completion acceptance.

6 Conclusions.

## 8.3 Main Acceptance Activities

**8.3.1** Review related documents and confirm whether all conditions required in Section 8.1 are met.

**8.3.2** Check the treatment of pending issues in special acceptance and the completion situation of the winding-up works, the construction and completion acceptance plan for the outstanding works, and assess their impacts on safe operation of the hydropower project.

## 8.4 Acceptance Procedures, Methods and Results

**8.4.1** After the acceptance presiding organization confirms that all special acceptances have been completed and the pending issues have been solved properly, the acceptance committee holds the completion acceptance meeting.

**8.4.2** The completion acceptance meeting includes the following:

1 Listen to the report from the project owner on the project construction.

2 Listen to the opinion of each presiding organization on completion acceptance.

3 Discuss to complete the completion acceptance certificate.

**8.4.3** The acceptance presiding organization is responsible for submitting the summary report of the completion acceptance, acceptance certificate and relevant documents to the provincial energy authority.

**8.4.4** The provincial energy authority issue the completion acceptance certificate (approval document) of the hydropower project, and copy to National Energy Administration.

# Appendix A Acceptance Program

## 1 Project Overview

1) Project name, geographical location, main structure name, scale, main indicators of comprehensive economic benefits (design parameters), etc.

2) Brief introduction to the project owner, and the full name of relevant organizations involving design, construction, supervision, equipment manufacture, safety monitoring, production and operation, resettlement implementation in the reservoir area, and their main tasks.

## 2 Main Acceptance Basis

Relevant approvals of the project, relevant state regulations and technical documents.

## 3 Acceptance Organization

1) Acceptance organization leadership.

A list of the chairman, deputy chairmen and members of the acceptance committee.

2) Acceptance committee office and its responsibilities.

The daily office of the acceptance committee is determined, which is responsible for the organization and arrangement of various activities of the acceptance committee and the expert team and the daily work.

3) Acceptance participants.

The participants of the acceptance meeting are determined, including the members of the acceptance committee and the project participants.

## 4 Acceptance Conditions and Basic Documents

The acceptance conditions and the requirements for the documents as acceptance basis shall be specified.

## 5 Acceptance Scope and Content

The scope, content and requirements of acceptance are specified.

## 6 Acceptance Procedures and Plans

The procedures, master plan and schedule of acceptance are specified.

## 7 Treatment of Problems in Acceptance

Principles and methods for dealing with the problems such as disputes in acceptance are specified.

## 8 Relevant Attachments

# Appendix B Acceptance Certificate

## 1 Foreword

Acceptance basis, acceptance presiding organizations and acceptance participants, and briefing of acceptance meetings.

## 2 Project Overview

1) Project name, geographical location, scale, comprehensive economic benefits (specific design indicators), water retaining structure, water release structure, water level, reservoir capacity, reservoir area, etc.

2) Full name of relevant organizations involving design, construction, supervision, equipment manufacture, safety monitoring, production and operation, resettlement implementation in the reservoir area, and their main tasks.

## 3 Acceptance Basis, Scope and Content

## 4 Project Construction and Its Assessment

1) Preparation completion and its assessment; planned implementation measures for uncompleted works; establishment and operation of safety monitoring systems of main structures, foundations and slopes, and hydraulic steel structures and equipment; technical schemes and their function assessment; major design changes and their approval; main conclusions of safety appraisal; main conclusions of project quality supervision.

2) Tasks and their implementation of the resettlement, and staged special acceptance and its comments. (Applicable to the river closure acceptance and impoundment acceptance)

3) River closure (or impoundment) scheme and flood control measures. (Applicable to the river closure acceptance and impoundment acceptance)

4) Each special acceptance and treatment of pending issues. (Applicable to the completion acceptance)

## 5 Existent Problems and Treatment Suggestions

Suggestions on dealing with the existent problems and pending issues.

## 6 Acceptance Conclusions

Clear conclusions on the approval of acceptance.

## 7 Attachments

Attachment 1: Signature form of committee members for ×× acceptance of ××× hydropower project (the discrepancy shall be signed by the person who raised the concern)

Attachment 2: Signature form of expert team members for ×× acceptance of ××× hydropower project

Attachment 3: Signature form of representatives of project participants for ×× acceptance of ××× hydropower project

Attachment 4: List of documents and information for ×× acceptance of ××× hydropower project

# Appendix C Basis Documents and Information for Acceptance

Table C Basis documents and information for acceptance

| Documents | Closure acceptance | Impoundment acceptance | Turbine-generator unit start-up acceptance | Special works acceptance | Hydropower complex acceptance | Completion acceptance |
|---|---|---|---|---|---|---|
| I Submittals | | | | | | |
| 1 Project construction report | √ | √ | √ | √ | √ | √ |
| 2 Supervision report | √ | √ | √ | √ | √ | |
| 3 Design report | √ | √ | √ | √ | √ | |
| 4 Construction report | √ | √ | √ | √ | √ | |
| 5 Production preparation and operation reports | √ | √ | √ | √ | √ | √ |
| 6 Comments on stage acceptance for resettlement | √ | √ | | | | √ |
| 7 Conclusions of project safety appraisal | ○ | √ | | √ | √ | |
| 8 Quality supervision report | √ | √ | √ | √ | √ | |
| 9 Emergency response plan | | √ | | | | √ |
| II Documents for reference | | | | | | |
| 1 Relevant design documents and contract documents | √ | √ | √ | √ | √ | √ |
| 2 Certificate for staged acceptance and special works acceptance | | √ | √ | √ | √ | √ |

**Table C** *(continued)*

| Documents | Closure accep-tance | Impound-ment accep-tance | Turbine-generator unit start-up accep-tance | Special works accep-tance | Hydro-power complex accep-tance | Comp-letion accep-tance |
|---|---|---|---|---|---|---|
| 3 List of completed and uncompleted items of project to be accepted | √ | √ | √ | √ | √ | √ |
| 4 Acceptance confi-rmation of supervision engineer | √ | √ | √ | √ | √ | |
| 5 Project safety appr-aisal report and its atta-chments | ○ | √ | | √ | √ | |
| 6 Major issue consulting report | ○ | ○ | ○ | ○ | ○ | ○ |

NOTE "√" denotes mandatory, and "○" denotes optional as needed.

# Explanation of Wording in This Specification

**1** Words used for different degrees of strictness are explained as follows in order to mark the differences in executing the requirements in this specification.

**1)** Words denoting a very strict or mandatory requirement:

"Must" is used for affirmation; "must not" for negation.

**2)** Words denoting a strict requirement under normal conditions:

"Shall" is used for affirmation; "shall not" for negation.

**3)** Words denoting a permission of a slight choice or an indication of the most suitable choice when conditions permit:

"Should" is used for affirmation; "should not" for negation.

**4)** "May" is used to express the option available, sometimes with the conditional permit.

**2** "Shall meet the requirements of..." or "shall comply with..." is used in this specification to indicate that it is necessary to comply with the requirements stipulated in other relative standards and codes.